Hydroponics

The Ultimate guide on how to grow your vegetables and fruits Step-by-step

[Green Bray]

Legal & Disclaimer

The information contained in this book and its contents is not designed to replace or take the place of any form of medical or professional advice; and is not meant to replace the need for independent medical, financial, legal or other professional advice or services, as may be required. The content and information in this book has been provided for educational and entertainment purposes only.

The content and information contained in this book has been compiled from sources deemed reliable, and it is accurate to the best of the Author's knowledge, information and belief. However, the Author cannot guarantee its accuracy and validity and cannot be held liable for any errors and/or omissions. Further, changes are periodically made to this book as and when needed.

Where appropriate and/or necessary, you must consult a professional (including but not limited to your doctor, attorney, financial advisor or such other professional advisor) before using any of the suggested remedies, techniques, or information in this book.

Upon using the contents and information contained in this book, you agree to hold harmless the Author from and against any damages, costs, and expenses, including any legal fees potentially resulting from the application of any of the information provided by this book. This disclaimer applies to any loss, damages or injury caused by the use and application, whether directly or indirectly, of any advice or information presented, whether for breach of contract, tort, negligence, personal injury, criminal intent, or under any other cause of action.

You agree to accept all risks of using the information presented inside this book.

You agree that by continuing to read this book, where appropriate and/or necessary, you shall consult a professional (including but not limited to your doctor, attorney, or financial advisor or such other advisor as needed) before using any of the suggested remedies, techniques, or information in this book.

INTRODUCTION 5

CHAPTER 1: WHAT IS HYDROPONICS? 9

A Brief History of Hydroponics 11
Is Hydroponics worth the while? 13
Why Hydroponics? 15
How Plants Grow 17

CHAPTER 2: THE BASICS OF HYDROPONICS 20

The Basics: What is a hydroponic system? 20
Here is a list of benefits of hydroponic systems: 23
Seeds and Seedlings 25
Tips for Beginners 25

CHAPTER 3: DIFFERENT KINDS OF HYDROPONIC GARDENS 31

1. The Water Culture System 32
2. The Wick System 33
3. Nutrient Film Technique 33
4. The Flood and Drain System 34
5. Drip System (The Recovery / Non-Recovery System) 37
6. Aeroponic 38

CHAPTER 4: THE SCIENCE BEHIND HYDROPONICS 39

Choosing The Best Hydroponic System In Relation To Your Budget And Needs 43
Light 43
Supplements 44
Acquired Containers 44
Reused Containers 45
Least Expensive Way to Grow Hydroponics 45
Backing 46
Plants that occupy enormous environments. 51
Profound Root Veggies 52

CHAPTER 5: HOW HYDROPONICS GARDENING WORKS...53

CHAPTER 6: DIFFERENCE BETWEEN HYDROPONICS AND SOIL GARDENING 55

AREAS WHERE HYDROPONIC GARDENING IS BETTER 57
AREAS WHERE SOIL GARDENING IS BETTER..59

CHAPTER 7: ADVANTAGES AND DISADVANTAGES OF HYDROPONICS GARDENING...61

THE ADVANTAGES OF HYDROPONIC GARDENING 61
DISADVANTAGES OF HYDROPONIC GARDENING 64

INTRODUCTION

It is an understatement to say that without water, human existence is equated to nothing. That is, water is life. There would be no hydroponics to practice or creating a society of it. Water is an essential part of all living cells. This induces turgor pressure on cell walls of plants to prevent the leaves from wilting. And it holds nutrients and energy reserves all over the plant in the form of dissolved salts and sugars.

This book is about water, focusing on how it is being distributed, how to preserve its consistency, and how to supplement it with plant life's essential nutrients. Fire and water work together in nature so as to recharge the soil with nutrients. Woods turns to ash as trees burn in the forest. Wood ash is high in Potassium, one of the essential nutrients of the plant kingdom.

The organic matter in the soil is bio-decomposed into the basic nutrient salts on which the plants feed. The dropping rains continue to remove these salts once again, making them available for plants to consume from their roots. Everything in nature should be in perfect harmony if a plant is to be provided a well-balanced or adequately balanced diet. Forests must burn, cattle must feed, rains must fall, the wood must rot, and there must be bacteria in the soil and ready to go to work.

You will never ever see these ideal conditions happening regularly. Besides, the rainforests of the world might be the only remaining examples of near-perfect botanical conditions.

Now that we grasp the natural growing process better, we can see that hydroponics is all about enriching water with the very same nutritional salts found in nature. It's about creating and maintaining a perfectly balanced "nutrient solution" for your plants. Many hydroponic systems have a closed system that stores the nutrient solution. It helps protect it from evaporation and pollution into our atmosphere, just as the wastewater from untreated fertilized soil does.

This conservative approach to water management makes hydroponics the tool of choice in drought-stricken areas around the world, and as a result, it is quickly becoming recognized as "Earth Friendly Farming." Since you will be learning the art and science of "climate farming," it is a good idea to know what your local water contains. Contact your local water company and request their analysis of water quality.

If your water comes from a well, you'll most likely have to take it out to your own laboratory for study. The relative "hardness" or "softness" is the most important factor influencing water quality. Hard water means that there is a lot of dissolved mineral content, mostly calcium carbonate, and is often seen as a scale on hot water pipes. Soft water in dissolved solids is usually very pure, or weak. Distilled (or deionized) water, or water going through a tank for reverse osmosis, is all called liquid. Most of the marketable hydroponic products are made for soft water. If you have hard water, though, there are some nutrient items that are also designed for hard water.

Chapter 1: What is Hydroponics?

The word hydroponic means working "with water." In simple terms, it is the science of growing plants without dirt or soil as a medium.

Plants need nutrients to grow and to anchor them for support. Plants usually get these from the soil in traditional gardening. However, in hydroponics, you can give plants exactly what they need, in the particular amount needed and when they need it. You also need to provide the plants with extra support, but it is quite easy. Enriched water can easily provide all the nutrients required by the plants with very little extra work. In fact, this is easier to do in water than in soil.

The plants receive pH-adjusted nutrient solutions. The roots absorb nutrients more efficiently in a highly soluble form. It takes very little effort for the roots to absorb the needed nutrients in order to grow. Even if the soil is organic and rich in nutrients, the plant will expend too much energy searching for and extracting those nutrients.

Hydroponic plants grow easily and produce more because the force it takes for the roots to wheedle out nutrients in the soil goes instead on vegetative growth and bearing fruit or flowers. Thus, it's more effective as a means of production.

Hydroponics may sound complicated, but you can think about it in simple terms, such as a simple system of growing plants in a water bucket. It is very easy and even a child can take it up as a hobby. You do not even need many instruments and apparatus to start hydroponic gardening. An average home hydroponic unit will just require a few things: a reservoir, a growing tray or medium, a timer, submersible pump, an air stone and air pump. Light can be natural or artificial. That's all that is required and although that may sound complex, it isn't.

The growing medium is the passive material in hydroponics. It is the substance where the roots grow. Unlike soil, it does not provide any kind of nutrition on its own. Some growing mediums include perlite, rock wool, coconut fiber, vermiculite, sand, and gravel.

Because the growing medium is inert, you can control the nutrients that the plant receives by adjusting the pH levels and strength of the nutrient solutions. You also manage the feeding and watering cycles. With technology, the potential to have a high-tech hydro system is not impossible. All aspects of this type of gardening are therefore easy, automated and controlled. The only limit is your budget and your imagination.

A Brief History of Hydroponics

The American customer, even with all its benefits, is sometimes suspicious of hydroponically grown food.

Hydroponic products were of poor quality, admittedly, many years ago, and this association still persists for some people. The old narrative, however, is changing rapidly as hydroponic produce has grown into a premium product of superior quality. Nonetheless, modern-day hydroponic cultivation has become so successful, NASA itself has developed an innovative hydroponic system for outer space use. Although hydroponics may appear to be a recent invention, its history can be traced back to the dawn of civilization.

Once grown hydroponically, their roots are bathed or sprayed directly with water-dissolved nutrients. Since they no longer need to search for food, they can redirect most of their energy into the production of foliage, flowers, fruits, and vegetables. Hydroponically grown plants are better because they provide a well-balanced "diet." They are more robust because little effort is devoted to seeking water and nutrients. As a result, hydroponically grown products are usually larger, more tasteful, and more nutritious than soil-grown produce. A clean, sterile medium such as sand, gravel, rocks, coco fiber, or Rockwool (or combination of each) may be used to give the physical support that soil would normally provide.

There is no medium, in the case of aeroponics, with which plants receive physical support from baskets and even wires suspended from the roof. At Epcot, plants are rotated through a chamber with a fine mist of water and nutrients supplying their roots. The extra Oxygen entering the roots significantly increases the metabolism of the plant.

Is Hydroponics worth the while?

Gardeners enjoy hydroponics because it's possible to grow almost anything, and there's little or no backbreaking work: no tilling, raking or hoeing. There are no pulling of weeds and no spraying of toxic pesticides. Few moles or cutworms consume the roots, and most insects leave their plants clean and healthy alone.

Hydroponics is suitable for the homeowner or tenant hobbyists who have no time or space for full-time gardening of the land. In late spring and summer, you can place your portable hydroponic unit outside on a porch or balcony where natural sunlight helps to produce tremendous yields from lettuce, to cucumbers, to zinnias. The unit can be moved anywhere inside the home in winter, even into the basement, where your plants will thrive and continue to be produced under artificial light.

Plants love to grow through hydroponics because their roots don't have to pass through thick, chunky soil to fight for nutrients. Alternatively, one hydroponic system equally distributes nutrients to each vine. Plants also need oxygen to breathe, and a porous expanding aggregate, unlike dirt, allows air to circulate around them freely. And everything is growing fast and beautifully.

Hydroponic plants grow quicker, mature earlier and give up the yield of soil-grown plants by up to ten times. This washed and pampered plants produce high nutritional value fruits and vegetables, with outstanding flavor. Many of them, particularly hydroponic tomatoes and cucumbers, are sold at far higher prices than ordinary vegetables in the gourmet sections of supermarkets. The point here is that the same vegetables can be grown for considerably less money than the pulpy supermarket variety costs to buy.

Why Hydroponics?

Have you noticed lately that vegetables in supermarkets are missing something? It's all flavor. As with many industrial foods, the taste was substituted for consumer convenience. Large-scale farming and marketing do, of course, provide the world's burgeoning population with vast quantities of food, but it is important to remember that quality suffers whenever quantity is stressed. As a result, your meals ' flavor and nutritional value are reduced.

The varieties of seeds produced for "agribusiness" are one major reason for these declines. These seeds are selected for fast growth and high yield. The resulting vegetables and fruits have rugged skins for processing, storage, and shipping devices. Flavor and price are concerns secondary. Also, many vegetables are harvested unripe, particularly tomatoes, to ensure safe shipment and longer shelf life in the market. In fact, attempts are now being made to develop a hybrid square tomato that fits in packages.

More often than not, cities and villages grew up in frontier days, where farmers till the soil. They were good farmers and had the best soil to choose from. These towns and villages are our present-day cities, still expanding and still engulfing valuable agricultural land. As prime agricultural land disappears, as the costs of growers continue to rise, as transport costs increase in parallel with energy supplies, and as supermarket boards of directors become increasingly concerned with profit margins, we will see our food costs rise to the point of absurdity. World War II's Victory Gardens have been planted to raise unavailable food, and it seems realistic to say millions of people will use hydroponics in the near future to supply themselves with affordable vegetables and herbs of a quality that stores cannot match.

How Plants Grow

Many hydroponics books, complete with illustrations, give the reader a crash course on plant biology. Relating biology specifically to the hydroponics and the nutrients that help plants grow seems to make more sense.

The growing plant is a natural workshop which constructs organic matter in the form of roots, stems, leaves, fruit, and seed. Less than ninety-seven percent of this matter is provided by air and water, while the rest comes from plant nutrients. No organic substance can be taken up by a plant; instead, it absorbs inorganic mineral salts. That is, the vegetable kingdom feeds directly upon the kingdom of minerals.

There is no dispute, however, between organic gardening and hydroponics. In organic gardening, however, the difference is that it is the soil that is fed with dead plant and animal matter, not the plant. Soil acts as a natural fertilizer factory that works with its soil bacteria in league with weathering to operate on those organic substances. It breaks down these substances into their inorganic composites (chemicals, if you like), so they can be fed upon by the plants.

There is no soil in hydroponics, and the plants are fed directly with the same minerals to produce healthy organic soil. The plant doesn't know if its mineral food was made by man or nature or a particular care pattern. However, it does care that it is well fed, and a nitrate is a nitrate, whether it comes from a solution to the nutrient or a dead mouse.

To grow, a plant utilizes two basic processes. The first, osmosis takes over the roots from water and minerals. The second, photosynthesis, turns the water and minerals into plant tissue using light and the environment. To breathe, roots also need oxygen, and this is one of the reasons that hydroponics functions so well. The loose, chunky growing medium hydroponic, the aggregate, as it's called, allows plenty of air to reach the roots. Natural soil, on the other hand, often requires much work and time to ensure satisfactory aeration.

Chapter 2: The Basics of Hydroponics

The Basics: What is a hydroponic system?

Hydroponics can be described by simply saying that is a process of growing plants with water and nutrients without the use of soil. The water is given to the roots of the plants that are being grown. The plant roots may hang in the nutrient solution, misted, enclosed inside of a container, or a trough that is filled with a soil substitute. The substitute can consist of materials like sand, perlite, sawdust, pebbles, wood chips, or rockwool. Any substitute being used will need to provide great water holding capabilities, yet be porous enough for gas exchange. Between watering the plants, it will become a storage area for water and nutrients for the root system. The plant roots grow in the substitute in order to secure the plant inside the container or the trough.

There are many different methods of delivering water to the root of the plant. For the growth inside containers, each of the plants will need to be provided with an emitter for the water from an irrigation system. Water can be channeled to a row of plants inside of a trough like the nutrient film method. A large tray of specific plants can be watered from blow by filling the tray with water, and then allowing it to drain all of the excess water. This is called flood irrigation. Water is then recycled within the nutrient film method and flood systems. It is harder to recycle using a drip irrigation system and it requires extra equipment like water sterilizer and fertilizer monitoring, as well as adjustment equipment.

In combination with a greenhouse, hydroponics is a technology and is inexpensive. Hydroponics does not require a lot of knowledge and can be done with the most basic information; however, it can be done a much larger scale. It will depend on the land you have available and the time. It will also depend on the purpose. For example, a person only wanting to grow herbs will have much less work than one who wants to grow a years' worth of vegetables for a family of five. Since the regulating of the aerial and the root environment is crucial, hydroponics is usual performed inside of a shelter. The enclosures are used to offer control over air and root temperatures, light, water, nutrition, and the climate.

There are different types of controlled environments. Each component of the environment agriculture, also called CEA, is equally crucial to the process. Not every hydroponic system is cost effective. If attention is not balanced from the structure to the environment, the system will prove to be less productive than have planned. Therefore, it is extremely important to pay equal attention to every aspect of the hydroponic system.

Here is a list of benefits of hydroponic systems:

Offers the ability to produce higher yields than soil-based agriculture.

Allows food to grow anywhere that does not support soil crops.

Overcome seasonal limitations.

They are portable.

Harvesting is extremely easy.

Eliminates the need for pesticides.

Is a learning and fun experience that the whole family can be involved in.

Hydroponically grown plants like basil, lettuce, and other plants can be packaged and given or sold while they are still alive, which prolongs their shelf life.

Solution hydroponics does not require any disposal of solids.

Hydroponics will allow a greater control over the plant root zone.

Hydroponics is often the best way to produce a crop method in different remote areas that do not have suitable soil.

Solution hydroponics offers visible roots.

They are great for teaching, as well as research.

Solution hydroponics keeps your plants safe from soil-based plant diseases.

You will not have to deal with weeds.

The crops are not contaminated by the soil.

They cost less than soil-based gardens.

Plants tend to be healthy and grow larger, as well as faster.

Temperature fluctuations, over or under watering, wind damage, lack of light, and excess sunlight can be controlled.

Shade cloth is easy to install and removed.

It is flexible. You can do it as a hobby or professionally.

Seeds and Seedlings

Seedlings that are transferred to a hydroponic system grow inside the channel and can be grown with the roots exposed or even in root cubes. It is the very best way to grow them using a water culture system. Sow your seeds in quartz sand, cellulose fiber, vermiculite, or perlite. These cubes will offer weight and help support your plant. Offer water to your seeds and then cover them with paper towels or cheesecloth that is wet until they germinate. You will then need to remove the covering and thin your plants. Moisten them using a solution that is infused with the right amount of hydrogen peroxide, rather than just water.

Tips for Beginners

Growing plants using hydroponics is a huge step for many individuals and can be a large learning curve.

In order to help you, here are signs of deficiencies in order to catch them before your plants cannot recover.

There is no dispute, however, between organic gardening and hydroponics. In organic gardening, however, the difference is that it is the soil that is fed with dead plant and animal matter, not the plant. Soil acts as a natural fertilizer factory that works with its soil bacteria in league with weathering to operate on those organic substances. It breaks down these substances into their inorganic composites (chemicals, if you like), so they can be fed upon by the plants.

There is no soil in hydroponics, and the plants are fed directly with the same minerals to produce healthy organic soil. The plant doesn't know if its mineral food was made by man or nature or a particular care pattern. However, it does care that it is well fed, and a nitrate is a nitrate, whether it comes from a solution to the nutrient or a dead mouse.

To grow, a plant utilizes two basic processes. The first, osmosis takes over the roots from water and minerals. The second, photosynthesis, turns the water and minerals into plant tissue using light and the environment. To breathe, roots also need oxygen, and this is one of the reasons that hydroponics functions so well. The loose, chunky growing medium hydroponic, the aggregate, as it's called, allows plenty of air to reach the roots. Natural soil, on the other hand, often requires much work and time to ensure satisfactory aeration.

SYMPTOMS	DEFICIENCY
The entire plant is light green. The lower leaves are turning yellow. The growth seems to be stunted.	Nitrogen
The entire plant is almost a bluish, green color. It may develop a red or even purplish cast. The lower	Phosphorous

leaves might be yellow. They dry to a greenish, brown to black. The growth may be stunted.	
The plant leaves may be a papery appearance. The dead areas are present along the edges of leaves. The growth might be stunted.	Potassium
The lower plant leaves will turn yellow at the tips and the margin between the veins. The lower leaves may be wilting.	Magnesium
The young stems and the new plant leaves die.	Calcium
The leaf tissue between the plant veins is lighter in color. It may be yellowed and	Zinc

papery.	
The leaf tissue will appear yellow and the veins will still be green	Iron
The plant's leaf's will be dark green or blue around the edges. The leaf edges will curl upward. Young leaves will wilt.	Copper
The young leaves will turn pale green and the older leaves will stay green. The plant will be stunted.	Sulfur
The growth will be stunted. The lower leaves will have a checkered pattern of green and yellow.	Manganese
The leaves will be stunted, a pale green, and they will be	Molybdenum

malformed.	
The young leaves will be scorched at the ends and the margins.	Boron

Note: In order to fix the deficiency, you will need to mix up a solution in a spritzing bottle. You will need to offer the solution to the plant on a consistent basis until there is a balance. In some cases, it may be best to switch the type of hydroponic fertilizer that you are using. Make sure to watch your plants to catch the deficiencies early so that you can keep them alive and ensure that they thrive.

Chapter 3: Different Kinds of Hydroponic Gardens

There are hundreds of methods of hydroponic gardening. However, all these are combinations or variations of six basic types:

Water Culture

Wicks

Nutrient Film Techniques

Flood and Drains

Drips

Aeroponics

The following paragraphs give descriptions of the basic hydroponic systems and details how each of them works.

1. The Water Culture System

The Water Culture System is the simplest form of active hydroponics. The plants are commonly grown on a medium made of Styrofoam and they grow directly from the nutrient solution. This system uses an air pump to supply oxygen to the plants' roots.

The Water Culture System is the best choice for growing water-loving plants such as lettuce. However, this system is not suitable for many long term or large plants, and these will not thrive using this system.

It is not expensive to make this type of hydroponic system. You can use an old aquarium or water container. It is the ideal set-up for a classroom. This makes it a popular choice for teachers and students.

2. The Wick System

Wicking is the simplest form of passive hydroponics. Passive means there are no moving parts in the system. The nutrient solution comes up through the wick from the reservoir and feeds the growing medium through this wick.

The growing mediums used for this system are coconut fiber, pro-mix, vermiculite and perlite. It is an effective system for small plants because large plants tend to draw up the nutrient solution faster than the wicks can supply them.

3. Nutrient Film Technique

The Nutrient Film Technique or NFT is the most prevalent type of hydroponics. It is probably the cheapest and easiest to create. The benefit of this system is that no soil is used. The roots of the plants are suspended directly in water and the nutrient solution is pumped into the water that covers the roots and drained back into a reservoir.

There is no need for a timer and you do not have to replace the growing medium after every change of crop. The NFT usually makes use of a small plastic basket that has been designed to let the roots dangle into the nutrient solution. The only drawback is that when power outages and pump failures occur, the flow of solution is interrupted and the roots tend to dry out easily.

4. The Flood and Drain System

This is known as the "Ebb and Flow." It works by flooding the growing medium or tray with the nutrient solution and draining it back to the reservoir. This action is achieved by a submersible pump, which is connected to a pre-set timer.

The timer will trigger the pump to siphon the nutrient solution onto the tray. After this action, the timer will also shut the pump off so that the solution will ebb back. The gardener will set the timer to turn on several times during the day – the frequency will be dependent on several factors

- Type of plant
- Size of plant
- **Temperature**
- **Humidity**
- Growing medium

The grow tray can be filled with different growing mediums. The most popular choices are rockwool, perlite, gravel, coconut fiber and grow rocks. Most people use individual pots as trays.

There is no dispute, however, between organic gardening and hydroponics. In organic gardening, however, the difference is that it is the soil that is fed with dead plant and animal matter, not the plant. Soil acts as a natural fertilizer factory that works with its soil bacteria in league with weathering to operate on those organic substances. It breaks down these substances into their inorganic composites (chemicals, if you like), so they can be fed upon by the plants.

There is no soil in hydroponics, and the plants are fed directly with the same minerals to produce healthy organic soil. The plant doesn't know if its mineral food was made by man or nature or a particular care pattern. However, it does care that it is well fed, and a nitrate is a nitrate, whether it comes from a solution to the nutrient or a dead mouse.

To grow, a plant utilizes two basic processes. The first, osmosis takes over the roots from water and minerals. The second, photosynthesis, turns the water and minerals into plant tissue using light and the environment. To breathe, roots also need oxygen, and this is one of the reasons that hydroponics functions so well. The loose, chunky growing medium hydroponic, the aggregate, as it's called, allows plenty of air to reach the roots. Natural soil, on the other hand, often requires much work and time to ensure satisfactory aeration.

The main challenge with the "Ebb and Flow" system is the susceptibility to power outages, pump failures, and timer failures. Some mediums like gravel and grow rocks will not hold the nutrient solution well enough so the roots will dry out quickly when the cycle is interrupted. It is better to use rockwool, coconut fiber, and pro-mix as they retain more water.

5. Drip System (The Recovery / Non-Recovery System)

The Drip System is the most widely used hydroponic system. It is set-up with a timer, a submerged pump, and a grow tray. The timer is set to turn the pump on to allow the nutrient solution to drip off directly onto the plants through a tiny drip line.

There are two kinds of Drip Systems: Recovery and Non-Recovery. In a Recovery Drip, the surplus nutrient solution that flows down is collected in a reservoir and re-used.

The Recovery System requires more maintenance in recycling the solution back to the reservoir and the pH and strength of the nutrient solution needs to be preserved. This requires periodic testing and adjusting so that pH and strength levels do not shift. On the other hand, the Non-Recovery System needs less maintenance, as the solution is not re-used.

6. Aeroponic

This is the one of the high-tech systems in hydroponic gardening. Like the NFT, the growing medium for the Aeroponic system is air. The plants' roots are misted with the nutrient solution every few minutes.

A timer triggers the misting pump, similar to on the other hydroponic systems. The only difference is that there is a shorter cycle for the pump.

It is a quite delicate and complicated system. There should be no interruption to misting cycles. Otherwise, the roots will dry out quickly.

Chapter 4: The Science behind Hydroponics

Before we can investigate how hydroponics works, we should initially see how plants themselves work. As a rule, plants need next to no to develop. They can subsist on a straightforward mix of water, daylight, carbon dioxide and mineral supplements from the dirt.

The photosynthesis procedure necessitates that the plant approaches certain minerals, particularly nitrogen, phosphorus and potassium. These supplements can be normally happening in soil and are found in most business manures. Notice that the dirt itself isn't required for plant development: the plant essentially needs the minerals from the dirt.

This is the essential reason behind hydroponics - every one of the components required for plant development are equivalent to with conventional soil-based cultivating. Hydroponics basically removes the dirt necessities.

There are a few distinct sorts of hydroponic frameworks, however each depends on a similar beginning idea. Here, we'll inspect each type, find how and why it's utilized and see which sorts of plants react best to every strategy.

Back and forth movement Systems require a medium, for example, perlite, which fills no need other than to give solidness to the plant's underlying foundations. The plant gets no supplements from the medium itself. Back and forth movement frameworks incorporate a plate in which the plant is set in a medium; underneath the plate in a different holder is a store containing water and mineral arrangements. The water from the store is intermittently siphoned up into the plate. This floods the plate and enables the plants to ingest water and supplements. Bit by bit, the water depletes once more into the repository because of gravity. Back and forth movement frameworks work best with little plants like herbs and are normally utilized in smaller hydroponic arrangements, for example, those in the home.

Supplement Film Technique (NFT) is a water-based framework that requires no dirt or mediums. They're assembled utilizing wooden channels, which bolster polyethylene film liners. Plants, for example, tomatoes and cucumbers are set on the channels, and the supplement improved water is siphoned to the high finish of each channel. The channels slant down, and water is gathered toward the conclusion to be siphoned back through the framework and reused. Just plants with enormous built up root frameworks will work with this procedure.

Trickle Systems are set up indistinguishably from a back and forth movement framework, despite the fact that rather than water being siphoned through one enormous cylinder, it's siphoned through numerous little cylinders and channels onto the highest point of the plants. This framework is perfect for plants that don't yet have a created root framework, and like a recurring pattern framework, works best with smaller plants.

Aeroponics is another water-based framework, which, as NFT, requires no medium. Plants are suspended on a plate, with their foundations uninhibitedly dangling underneath. The whole plate is set into a crate that has a limited quantity of water and supplement arrangement in the base. A siphon framework is utilized to draw the water up, where it's showered in a fine fog onto the whole plant and root in a consistent way. This framework is the hardest to set up and oversee, yet it has incredible potential for enormous business employments.

Wick Systems are like rhythmic movement frameworks in that they're medium-based. Plants are set into a plate loaded up with a medium, for example, perlite or rockwool. At the base of each root, a nylon rope is set, which is permitted to dangle openly, reaching out past the base of the plate. The whole plate is then set over a repository. The nylon ropes retain the water and supplements, wicking them up to the plant's underlying foundations. This framework is alluring in light of the fact that it requires no siphons or other gear to be bought.

Choosing The Best Hydroponic System In Relation To Your Budget And Needs

Light

For unquestionably the least expensive technique for developing put your hydroponic framework outside and exploit the daylight.

Keep up the separation between the light and plants by raising it as they develop. Shop lights are cheap to purchase, keep up and work. They give satisfactory lighting to a wide range of greens, however on the off chance that you need to develop whatever blossoms or natural products you'll have to get a fluorescent develop light.

Supplements

Your hydroponic plants are absolutely reliant on the supplements you give them, since they can't send their underlying foundations out into the dirt to discover the things they need. Hydroponics providers offer a huge scope of supplements, some of which are specific and some that aren't. These can wind up costing a considerable amount and aren't vital for a fundamental hydroponic arrangement. Make your very own supplement arrangement by including two teaspoons of water solvent compost and one teaspoon of Epsom salts to every gallon of water. Blend until the solids are broken down and pour it in.

Acquired Containers

Plastic stockpiling receptacles, for example, the sort sold in most markdown stores, are modest and simple to discover. These have the upside of coming in a wide range of sizes so you can pick the one that best suits your needs and space. Purchase the dark ones to impede the development of green growth, since green growth can stop up your framework just as ransack your plants of the supplements they need. Five-gallon basins are likewise simple to discover and function admirably for bigger plants, for example, tomatoes and eggplant.

Reused Containers

A compartment for an essential hydroponic framework doesn't need to be extravagant, costly or even a specific size. One was to set aside cash when setting up your hydroponics framework is to utilize reused materials to hold your plants. Froth holders used to ship fish will function admirably, as will an old cooler, a youngster's swimming pool or a straightforward plastic can. When picking a holder, ensure it doesn't spill and that it hasn't had anything unsafe in it like weed executioner or paint, since these can corrupt or slaughter your plants.

Least Expensive Way to Grow Hydroponics

A hydroponic framework can give an all year wellspring of nourishment for you and your family without requiring a significant venture. In case you're on a spending limit, attempt a profound water culture framework, an arrangement that is fundamentally a little lake with the plants suspended above it. Start with one holder and afterward include more as your spending grants. Regardless of whether you need to purchase everything this is a modest method to develop, and in the event that you are clever you can get moving for all intents and purposes nothing.

Backing

Plants in hydroponic frameworks need some sort of help to hold them set up. Net pots are promptly accessible, economical and reusable. To spare considerably more, jab a lot of gaps in plastic nourishment grade compartments and use them to hold your plants in the hydroponic framework. Keep plants upstanding inside the net pots utilizing disinfected pea rock or coconut coir, the two of which are economical. These substances are additionally reusable after sanitization, making them much increasingly reasonable.

The five best plants to develop in a hydroponic framework are:

Herbs

Bell Peppers

Strawberries

Spinach

Lettuce

Cultivators have discovered that these plants take to hydroponics like a duck to water. They're solid, quickly developing and don't take a ton of work to begin – every single incredible element that give another producer a little squirm room!

Presently how about we take a gander at every one of these somewhat closer:

Herbs in Hydroponics

Grow time: Varies by plant

Best pH: Varies by plant

Tip: Flush your developing medium about once per week to dispose of any additional supplements that your plants haven't (or won't) assimilate.

Variety choices: Name your top pick, and you'll discover guidelines for developing it!

Chime Peppers in Hydroponics

Chime peppers are a marginally further developed hydroponic plant. Try not to give them a chance to develop to their full tallness, rather, prune and squeeze plants at around 8 crawls to spike pepper development. Profound water culture or back and forth movement frameworks are best for peppers.

Grow time: About 90 days

Best pH: 6.0 to 6.5

Tip: Plan to give as long as 18 hours of light for these plants every day, and raise your light rack as the plants develop, keeping plants around 6 creeps from the lights.

Variety choices: Ace, California Wonder, Vidi, Yolo Wonder

Strawberries in Hydroponics

The most exceedingly terrible thing about strawberries is the way occasional they are. In the event that you don't get them locally when the yield is prepared, you're depending on trucked-in berries that start falling apart when they're picked. With hydroponics, you can have a prepared to-eat harvest of strawberries throughout the entire year. Collecting is super-helpful too - no twisting around! Strawberries appear to do best with a back and forth movement framework, however profound water culture or supplement film system can accomplish for a little harvest.

Grow time: About 60 days

Best pH: 5.5 to 6.2

Tip: Don't purchase strawberry seeds, which won't be berry-prepared for a considerable length of time.

Rather, you need to purchase cold-put away sprinters that are now at that stage.

Spinach in Hydroponics

Spinach develops rapidly in a hydroponic framework, especially when utilizing the Nutrient Film Technique or different strategies that keep the supplement arrangement profoundly oxygenated. You'll additionally use far less water than an in-the-ground garden. It's anything but difficult to begin these plants from seed and seven days in the wake of growing, move them into your framework.

Grow time: About 40 days

Best pH: 6.0 to 7.5

Tip: For better spinach, keep your develop temperatures between 65 degrees F and 72 degrees F. The lower temperatures may slow develop time, however.

Lettuce in Hydroponics

Lettuce (and most other verdant greens) ought to be your first plant to attempt with a hydroponic framework. These plants have a shallow root framework that matches their short over the ground tallness. That implies there's no compelling reason to tie stakes or set aides for the plant. Rather, you simply let them develop while consistently changing their supplement arrangement. In the long run, they will look adequate to eat, and you can!

Plants that occupy enormous environments.

In the event that space is constrained, it's ideal to maintain a strategic distance from squash, melons, pumpkins, corn and other huge plants. It doesn't imply that you can't develop these plants, yet in a limited region, it's harder to deal with plants and the yields are not in the same class as different spots where these plants have rooms to develop.

Profound Root Veggies

Once more, it's difficult to think about plants that need a great deal of profundity for root. So, this isn't prescribed for fledglings.

Potatoes, carrots, turnips fall into these sorts.

For root crops, you need a substrate with adequate length and high profundity to help the roots. What's more, these sorts of plants tend not to give as great outcomes as they are in the dirt.

In the event that you have a huge developing condition like a nursery, porch, you can set up a further developed framework and develop sizable plants, root veggies, and other difficult to-grow ones. That condition is ideal for you to attempt with any plant.

Chapter 5: How Hydroponics Gardening works

Biological decomposition in the soil breaks down the organic matter into nutrient salts that the plants will feed on. Water dissolves the salts that allow the uptake of all the nutrients by the plant roots. For a plant to receive a good balanced diet, everything inside the soil or solution must be in a perfect balance. The water quality can be an issue due to water that is alkalinity or even have a bad salt content. It will result in an imbalance of the nutrients and the plant will grow poorly. Softened water will contain amounts of sodium that proves to be harmful to the plant. Water that is high in salts should never be used. Salt that is great than .5 million or 320 parts per million will cause a bad imbalance.

Hydroponic plants will have the perfect balanced diet of food, as well as water that is delivered directly to the plant roots. With the proper exposure to light or sunlight; the plants grow and will develop with much less energy than it would to use a typical growing technique. A soil free medium like sand, fiber, coco fiber, or more can be used in order to anchor the plant roots. These mediums are porous. This means they hold the water and the air that is necessary for thriving plants. Air circulation around the plant leaves is crucial too. It will prevent plant diseases that are caused by stagnant, moist conditions. Indoor units often will have a small fan in order to circulate the air.

Chapter 6: Difference between Hydroponics and Soil Gardening

So, what makes hydroponics so different from traditional soil gardening methods? In this chapter, you are going to find out.

There is no dispute, however, between organic gardening and hydroponics. In organic gardening, however, the difference is that it is the soil that is fed with dead plant and animal matter, not the plant. Soil acts as a natural fertilizer factory that works with its soil bacteria in league with weathering to operate on those organic substances. It breaks down these substances into their inorganic composites (chemicals, if you like), so they can be fed upon by the plants.

There is no soil in hydroponics, and the plants are fed directly with the same minerals to produce healthy organic soil. The plant doesn't know if its mineral food was made by man or nature or a particular care pattern. However, it does care that it is well fed, and a nitrate is a nitrate, whether it comes from a solution to the nutrient or a dead mouse.

To grow, a plant utilizes two basic processes. The first, osmosis takes over the roots from water and minerals. The second, photosynthesis, turns the water and minerals into plant tissue using light and the environment. To breathe, roots also need oxygen, and this is one of the reasons that hydroponics functions so well. The loose, chunky growing medium hydroponic, the aggregate, as it's called, allows plenty of air to reach the roots. Natural soil, on the other hand, often requires much work and time to ensure satisfactory aeration.

Here are the advantages of each type of plant growing concept.

Areas Where Hydroponic Gardening is Better

Hydroponics prevents the overuse of fertilizer. Hydroponic plants are grown in a very controlled environment, where waste products are limited and less nutrient material is needed. The great thing about this control is that it allows less fertilizer to be used. This is especially beneficial for the humans and animals in the area, who will have less of a chance of drinking fertilizer-contaminated water.

Hydroponics make better use of space and location. You can grow an indoor hydroponics system anywhere that you have room, because it takes up so little space and everything that the plant needs can be provided by your system. Additionally, roots grown in the soil need room to spread out while plants grown hydroponically have root systems that do not need to spread out. This means that you can grow plants closer together and save space.

Hydroponics uses less water. You would think that a hydroponic growing system would use more water than traditional methods, but that is not true. When plants are grown hydroponically, they are given only the amount of water that they need. When you water plants that are in soil, some of the water is going to seep into the ground or leak out of the pot. It will also be evaporated. Therefore, the plants are actually receiving only a fraction of the water that you are providing. Hydroponic systems are much more efficient when it comes to water usage and you actually end up using 70 to 80 percent less water.

Hydroponics systems reduce weeds, pests, and diseases. When you use more traditional gardening methods, the soil that you grow in can be filled with diseases, pests, and other plant parts. Hydroponic systems do away with this problem almost entirely.

Hydroponic systems grow plants twice as fast as traditional methods. Do you know what that means? You can have more harvests each year.

Because hydroponic systems provide exactly what the plant needs without the plant having to hunt for it, the growing cycle is much more efficient.

Hydroponics makes it easier for you to tamper with the nutrients for growing. Every plant, like every person, is unique. Each type is going to thrive in certain environments and struggle in others. Hydroponics is fun in this way. You have the ability to adjust the amount of nutrients in the solution and adjust it until you have the perfect growth solution.

Areas Where Soil Gardening is Better

Soil gardening has a lower initial cost. While hydroponics systems vary in their initial cost, they can get quite expensive. Some of this cost will be offset by the lesser amount of water, fertilizer, and pesticides that you will need.

Soil gardening does not use electricity. In several hydroponics gardening techniques, you must use a light source. Additionally, some systems use electricity to create bubbles in the nutrient system to aerate the roots.

Soil gardening has a less risk of mold and bacteria growth. One disadvantage of hydroponics is that plants are grown in a very moist environment. This leaves the plants susceptible to growth of mold and sometimes dangerous bacteria if enough precautions are not taken.

Chapter 7: Advantages and Disadvantages of Hydroponics Gardening

The Advantages of Hydroponic Gardening

The reason plants grown hydroponically are often larger than their soil-grown counterparts has to do with the fact that absorbing nutrients through a liquid solution is much more efficient a process to the plant than it is with soil, the food is easier to digest, in a sense. Because of this ease of absorption hydroponic plants are usually upwards of fifty percent larger than soil grown plants! Some of the other advantages of hydroponic growing are:

- Crop yields are also significantly increased, and the nutritional content is often higher than the same plant grown in a more traditional manner. Not only are you getting more fruits and vegetables with hydroponic gardening, you're getting larger and healthier ones too!

- Creating a garden is possible even in spaces where there is no soil. That means that you can grow plants in living spaces like condominiums and apartments. It also opens up the possibility of converting multi-level buildings into full-fledged agriculture patches.

- It allows you to grow food in places that are traditionally not fit for agriculture, such as arid areas. Israel and Arizona have long adopted this technique of growing crops, which allows their citizens to enjoy home-grown food and also to expand their food market. It also allows remote and unreachable places that have no agricultural space, such as Bermuda, to grow their own crops. Areas like Alaska and Russia that experience short seasons for planting has also adopted hydroponics and incorporated them in greenhouses so that they can have better control of climate for their plants.

- Maintaining your hydroponic setup is easier than taking care of a soil garden. They tend to use less water, which may sound counter-intuitive since the plants are being grown in water. However, in this setup, the water is always being reused.

- Maintaining proper Ph levels is also easier. So too, is ensuring that your crops get the proper nutrition they both need and deserve.

- Pests are also often less of a concern, and those that do find their way around are more easily dealt with.

- Harvesting crops is also often a simpler process when growing hydroponically.

When looking at the above advantages, you might be thinking, "Where's the catch?" There has to be a downside, here right? Otherwise, hydroponic growing would be more widespread and commonplace. Hydroponics can seem complicated at first, the thick veil of technical jargon scares many people away before they have the chance to discover just how simple it can be to start growing plants hydroponically.

Disadvantages of Hydroponic Gardening

There are many benefits to hydroponic gardening. Lack of land, frequent supply of water and other environmental concerns can be conquered with hydroponics. With the right knowledge and proper techniques, it is a valuable system for commercial farmers and gardeners, but while hydroponics poses many benefits to modern gardening and farming, there are also disadvantages that come with it.

For one, the initial cost to set-up a properly designed and effective hydroponic system is high. In the long run, the conservation of water and nutrients may prove to be inexpensive but before you can enjoy those benefits, you need to set-up a hydroponic system with all the necessary equipment. Hydroponic equipment does not come cheaply. Additionally, technical knowledge and skills are required to maintain the equipment.

Other disadvantages of hydroponic garden systems are the following:

Compared to farming in large fields, hydroponic gardening may yield limited production.

Hydroponic gardening requires constant supervision. You need to be responsible and diligent because the plants depend on you for their survival.

If you do not have sufficient knowledge, you will have to go by trial and error. Some plants will flourish while others may fizzle. You should be prepared to encounter frustrations and disappointments.

Hydroponic gardens are interrupted and influenced by power outages and pump failures.

Because there is no soil to act as a buffer, the plant will wither and die rapidly once the system fails. If interruptions occur, the plants must be watered manually.

Should a water-borne organism or disease appear in your set-up, it will quickly spread and all of the plants will be affected. Hence, vegetative growth and production is disturbed.

As with any project, make sure you consider all aspects and count the costs before you decide to set-up your own hydroponic system.